Managing Neuropathic Pain
in the Diabetic Patient
Second Edition

Managing Neuropathic Pain in the Diabetic Patient
Second Edition

Andrew JM Boulton
Professor of Medicine
University of Manchester, UK, and
Diabetes Research Insititute
University of Miami, USA

Loretta Vileikyte
Senior Lecturer
University of Manchester, UK, and
University of Miami, USA

Published by Current Medicine Group, 236 Grays Inn Road, London, WC1X 8HL

www.currentmedicinegroup.com

Copyright © 2009 Current Medicine Group, a part of Springer Science+Business Media

First edition 2007
Second edition 2009
Reprinted 2009

All rights reserved. No part of this publication may be reproduced, stored in a retrieval system or transmitted in any form or by any means electronic, mechanical, photocopying, recording or otherwise without the prior written permission of the copyright holder.

ISBN 978-1-85873-433-0

British Library Cataloguing-in-Publication Data.
A catalogue record for this book is available from the British Library.

Although every effort has been made to ensure that drug doses and other information are presented accurately in this publication, the ultimate responsibility rests with the prescribing physician. Neither the publisher nor the authors can be held responsible for errors or for any consequences arising from the use of the information contained herein. Any product mentioned in this publication should be used in accordance with the prescribing information prepared by the manufacturers. No claims or endorsements are made for any drug or compound at present under clinical investigation.

Project editor: Alison Whitehouse
Designer: Keith Baker
Production: Marina Maher
Printed in Spain by Mcc Graphics

Contents

Author biographies	vii

1 Introduction to diabetic neuropathies — 1
Definition — 1
Epidemiology and natural history — 2
References — 4

2 Classification and clinical features — 7
Classification — 7
Sensory neuropathies — 8
Focal and multifocal neuropathies — 13
Autonomic neuropathy — 16
References — 18

3 Diagnosis and staging — 21
Patient history — 21
Patient description of pain — 23
Positive and negative sensory symptoms — 23
Assessment scales — 25
Clinical examination — 26
Quantitative sensory testing — 28
Electrophysiological testing — 29
Diagnosing cardiovascular autonomic neuropathy — 29
Staging of neuropathy — 31
When to refer to a neurologist — 31
References — 33

4 Management of neuropathic pain — 35
Metabolic control — 35
Pharmacological treatments — 36
Topical treatments — 41
Non-pharmacological treatments — 42
Treatment of cardiovascular autonomic neuropathy — 43
References — 46

Index — 49

Author biographies

Andrew JM Boulton, MD, FRCP, is Professor of Medicine at the Manchester Royal Infirmary and Visiting Professor at the University of Miami, USA. Among his many awards, his contribution to worldwide care of the diabetic foot was honoured by receiving the American Diabetic Association's Roger Pecoraro Lectureship and the European Association for the Study of Diabetes (EASD) Camillo Golgi prize, and he was the first recipient of the international award on diabetic foot research. In 2008 he was awarded the Harold Rifkin prize of the ADA for distinguished international service in the cause of diabetes and the first outstanding achievement award of the American Professional Wound Care Association. Professor Boulton was the founding Chairman of the Diabetic Foot Study Group and was previously Chairman of Postgraduate Education and then programme chair for the EASD. He is renowned worldwide as a leading educator and lecturer on neuropathy and the diabetic foot. He is the global chairman of the Diabetes Lower Extremity Research Group (DIALEX). Professor Boulton chaired the ADA's expert group on diabetic neuropathy that resulted in the April 2005 ADA statement on diabetic neuropathy, and was chair of the ADA Foot Council 2005–2007. He is a member of the editorial review board for *Diabetes/Metabolism: Research and Reviews*, *Acta Diabetologica*, *Diabetes Care*, *The Diabetic Foot* and *International Diabetes Monitor*. Professor Boulton has authored more than 350 peer-reviewed manuscripts and book chapters, mainly on diabetic neuropathy and foot complications.

Loretta Vileikyte, MD, PhD, is Senior Lecturer at the University of Manchester and visiting Assistant Professor of Medicine at the University of Miami, USA. Her main research interests lie in the psychosocial aspects of the lower limb complications of diabetes and she developed the first neuropathy-specific quality-of-life scale that has been used in several clinical trials of new medications for painful diabetic neuropathy. She previously held the first joint American and British Diabetes Association grant to develop a scale to assess patients' cognitive and emotional representations of neuropathy and is currently funded by the National Institutes of Health (NIH) to study the effects of stress on wound healing in diabetic neuropathic foot ulceration.

Chapter 1

Introduction to diabetic neuropathies

The neuropathies are among the most common long-term complications of diabetes mellitus, affecting up to half of patients. Diabetic neuropathies are characterised by a progressive loss of nerve fibres and may involve both autonomic and somatic divisions of the nervous system.

The early recognition and appropriate management of neuropathies in diabetic patients are important for a number of reasons:
- non-diabetic neuropathies may be present in patients with diabetes;
- many patients with diabetic neuropathy are asymptomatic and thus at risk of insensate damage to their feet;
- both symptoms and deficits in patients with diabetic neuropathy have an adverse effect on quality of life;
- autonomic neuropathy causes substantial morbidity and mortality, particularly if the cardiovascular system is involved; and
- effective symptomatic treatments are available.

Definition

In 1998, an international consensus group agreed the following simple definition for use in clinical practice: 'Diabetic neuropathy is the presence of symptoms and/or signs of peripheral nerve dysfunction in people with diabetes after the exclusion of other causes' [1].

The same group agreed that neuropathy cannot be diagnosed without a careful clinical examination, and that an absence of symptoms must not be equated with absence of neuropathy. The definition also conveys the importance of excluding non-diabetic causes, which account for around 10% of peripheral neuropathies in diabetic patients [2].

It is also widely accepted that diabetic neuropathy should not be diagnosed on the basis of one sign, symptom or test alone; a minimum of two abnormalities (to include a clinical or quantitative test) is recommended [3].

Care must be taken to distinguish between the diagnosis of early diabetic neuropathy and that of 'loss of protective sensation' (LOPS), which is used to identify the foot at risk of ulceration. Whereas two simple clinical tests may be used for the latter [4], the diagnosis of early neuropathy may require more investigation and the exclusion of non-diabetic causes.

Epidemiology and natural history

The epidemiology and natural history of diabetic neuropathies are poorly understood. This reflects variations in diagnostic criteria, biased patient selection in observational studies, the asymptomatic nature of many neuropathies and the large pool of patients with undiagnosed diabetes mellitus.

Incidence and prevalence

Despite the limitations noted above, it is known that diabetes mellitus is the leading cause of neuropathy in the western world [5], and that neuropathies are the most common long-term microvascular complication of diabetes [6]. Based on several large studies, the annual incidence of neuropathy in patients with type 2 diabetes is believed to be around 2% [7]. The prevalence of symptomatic neuropathy may be as high as 21% [8], while neuropathic deficits are found on examination in up to 50% of all patients with diabetes [9,10]. A recent German population-based study [11] reported a prevalence of neuropathy of 28% in the diabetic population, whilst also highlighting that neuropathy also occurs in impaired glucose tolerance (IGT; 13%) and that the non-diabetic population may have signs of neuropathy (7%) which needs to be taken into account in any future epidemiological studies. Indeed, the neuropathy of IGT affects predominantly small fibres, and neuropathic pain is common is these patients (see below [12]).

Risk factors

Diabetic neuropathy has a multifactorial origin, and numerous metabolic and vascular factors have been implicated in the pathogenesis of the disease. Of all the proposed mechanisms, however, the strongest evidence exists for chronic hyperglycaemia. Indeed, it is clear that the development and progression of neuropathy are related to glycaemic control in both type 1 and type 2 diabetes.

In the Diabetes Control and Complications Trial, the annual incidence of neuropathy in type 1 diabetes was around 2% in conventionally treated patients versus just 0.56% in individuals with intensive glycaemic control [13].

In the landmark UK Prospective Diabetes Study, the progression of neuropathy in type 2 diabetes was dependent on glycaemic control [14].

In addition, longitudinal data from follow-up studies suggest that the duration and severity of hyperglycaemia are related to the severity of neuropathy [15,16].

Recent studies in patients with IGT provide further insights into the role of glucose metabolism and the development of neuropathy. In a study of patients with idiopathic painful neuropathies, IGT was significantly more common than it is in the general population [17]. Interestingly, neuropathy associated with IGT is milder than that associated with diabetes, and small nerve fibre involvement may be the earliest detectable sign of neuropathy [18].

The large EuroDiab prospective study in Europe has demonstrated the importance of vascular risk factors in the pathogenesis of diabetic neuropathy, including cigarette smoking, history of cardiovascular disease, hypertension and hypercholesterolaemia [19–21], in addition to height, alcohol consumption and duration of diabetes.

Natural history

In contrast to the many published reports on the prevalence of diabetic neuropathy, there are few on the natural history of the condition. The disease course is one of gradual progression. Although symptoms may wax and wane, and even resolve entirely, this does not imply improvement of the underlying pathology. Typically, the disappearance of symptoms reflects a worsening of the condition, with the resulting insensate foot at risk of ulceration [22].

References

1. Boulton AJM, Gries FA, Jervell JA. Guidelines for the diagnosis and outpatient management of diabetic peripheral neuropathy. Diabet Med 1998; 15:508–514.
2. Dyck PJ, Kratz KM, Karnes JL, et al. The prevalence by staged severity of various types of diabetic neuropathy, retinopathy, and nephropathy in a population-based cohort: the Rochester Diabetic Neuropathy Study. Neurology 1993; 43:817–824.
3. Boulton AJM, Malik RA, Arezzo JC, et al. Diabetic somatic neuropathies. Diabetes Care 2004; 27:1458–1486.
4. Boulton AJM, Armstrong DG, Albert SF, et al. Comprehensive foot examination and risk assessment. Diabetes Care 2008;31:1679–1685.
5. Greene DA, Feldman EL, Stevens MJ, et al. Diabetic neuropathy. In: Diabetes Mellitus. Porte D, Sherwin R, Rifkin H (eds), East Norwalk, CT: Appleton Lange, 1995.
6. Vinik AI, Park TS, Stansberry KB, et al. Diabetic neuropathies. Diabetologia 2000; 43:957–973.
7. The Diabetes Control and Complications Trial (DCCT) Research Group. The effect of intensive treatment of diabetes on the development and progression of long-term complications in insulin-dependent diabetes mellitus. N Engl J Med 1993; 329:977–986.
8. Abbott CA, Carrington AL, Ashe H, et al. The North-West Diabetes Foot Care Study: incidence of and risk factors for new diabetic foot ulceration in a community-based cohort. Diabet Med 2002; 19:377–384.
9. Pirart J. Diabetes mellitus and its degenerative complications: a prospective study of 4,400 patients observed between 1947 and 1973. Diabetes Metab 1977; 1:168–188.
10. Young MJ, Boulton AJM, MacLeod AF, et al. A multicentre study of the prevalence of diabetic peripheral neuropathy in the United Kingdom hospital clinicpopulation. Diabetologia 1993; 36:150–154.
11. Ziegler D, Rathmann W, Dickhaus T, et al. Prevalence of polyneuropathy in pre-diabetes and diabetes is associated with abdominal obesity and macroangiopathy: the MONICA/KORA Augsburg Surveys S2 and S3. Diabetes Care 2008;31:464-469.
12. Singleton AR, Smith AG. Neuropathy associated with prediabetes: what is new in 2007? Curr Diabetes Rep 2007;7:420-424.
13. The Diabetes Control and Complications Trial (DCCT) Research Group. The effect of intensive diabetes therapy on the development and progression of neuropathy. Ann Intern Med 1995; 122:561–568.
14. UK Prospective Diabetes Study (UKPDS) Group. Intensive blood-glucose control with sulphonylureas or insulin compared with conventional treatment and risk of complications in patients with type 2 diabetes (UKPDS 33). Lancet 1998; 352:837–853.
15. Dyck PJ, Davies JL, Wilson DM, et al. Risk factors for severity of diabetic polyneuropathy: intensive longitudinal assessment of the Rochester Diabetic Neuropathy Study cohort. Diabetes Care 1999; 22:1479–1486.
16. Partanen J, Niskanen L, Lehtinen J, et al. Natural history of peripheral neuropathy in patients with non-insulin dependent diabetes. N Engl J Med 1995; 333:89–94.
17. Singleton JR, Smith AG, Bromberg MB. Painful sensory polyneuropathy associated with impaired glucose tolerance. Muscle Nerve 2001; 24:1225-1228.
18. Sumner CJ, Sheth S, Griffin JW, et al. The spectrum of neuropathy in diabetes and impaired glucose tolerance. Neurology 2003; 60:108–111.
19. Tesfaye S, Stevens LK, Stephenson JM, et al. Prevalence of diabetic peripheral neuropathy and its relation to glycaemic control and potential risk factors: the Eurodiab IDDM complication study. Diabetologia 1996; 39:1377–1386.
20. Adler AI, Boyko EJ, Ahroni JH, et al. Risk factors for diabetic peripheral sensory neuropathy: results of the Seattle Prospective Diabetic Foot Study. Diabetes Care 1997; 20:1162–1167.

21. Tesfaye S, Chaturvedi N, Eaton S, et al. Vascular risk factors and diabetic neuropathy. N Engl J Med 2005;352:341-350.
22. Boulton AJM, Malik RA. Diabetic neuropathy. Med Clin North Am 1998; 82:909–929.

Chapter 2

Classification and clinical features

Classification

Several classification systems for the neuropathies have been proposed. Some are based on presumed aetiology whereas others refer to topographical features or disease pathogenesis. However, the inter-relationship of aetiology, mechanisms and symptoms is complex and poorly understood, and, as such, this traditional classification is of little use in clinical practice (Figure 2.1).

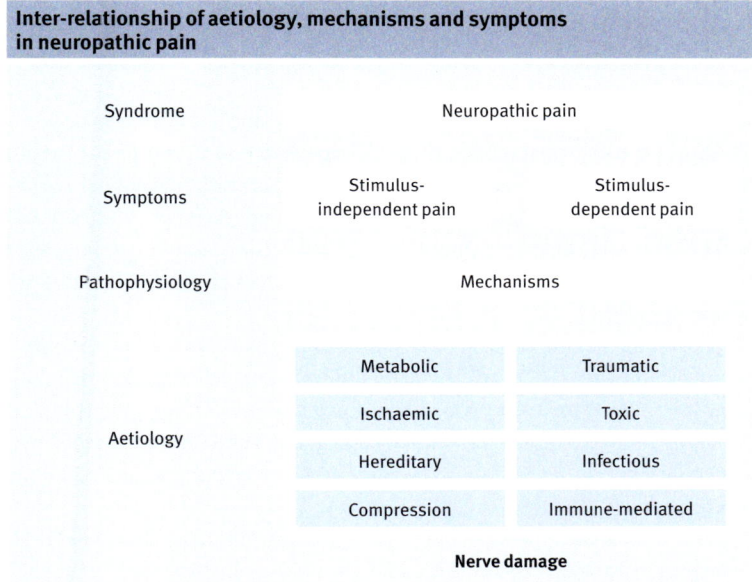

Figure 2.1. Inter-relationship of aetiology, mechanisms and symptoms in neuropathic pain.
Reproduced with permission from Woolf CJ, Mannion RJ. Neuropathic pain: aetiology, symptoms, mechanisms, and management. Lancet 1999; 353:1959–1964.

Until the underlying pathophysiology of diabetic neuropathies has been elucidated, the most useful systems for practising physicians are those based on clinical manifestations. Three classification systems are presented in Figure 2.2. This book follows the system used in the 2005 American Diabetes Association statement [1], which discusses diabetic neuropathies under three headings: sensory neuropathies; focal and multifocal neuropathies; and autonomic neuropathy.

The clinical features associated with each type of neuropathy are discussed in detail on the following pages. Pain terminology, as defined by the International Association for the Study of Pain, is summarised in Figure 2.3. The definition and classification of symptoms are described in more detail in Chapter 3.

Sensory neuropathies
Chronic sensorimotor distal symmetrical polyneuropathy
Chronic sensorimotor distal symmetrical polyneuropathy (DPN) is the most common form of diabetic neuropathy, being present in more than 10% of patients at the diagnosis of type 2 diabetes [2]. DPN occurs in both type 1 and type 2 diabetes and becomes more common with increasing age and duration of diabetes. In a large population survey, neuropathic symptoms affected 30% of type 1 diabetic patients and 36% and 40% of male and female type 2 diabetic patients, respectively [3,4]. However, 10% of men and 12% of women in the non-diabetic population reported similar symptoms.

Around half of patients with DPN experience symptoms, most often burning pain, electrical or stabbing sensations, paraesthesiae, hyperaesthesiae, and deep aching pain. Neuropathic pain tends to be intermittent and is typically worse at night. Symptoms are most commonly experienced in the feet and lower limbs but may extend to the hands and fingers in more severe cases.

DPN tends to have an insidious onset and many patients are truly asymptomatic, putting them at risk of foot ulceration and other late sequelae including Charcot's neuroarthropathy and even amputation [3,5]. Often, a neurological deficit is discovered by chance during a routine examination; patients may not volunteer symptoms, but on enquiry admit that their feet feel numb or dead [1].

Recently, unsteadiness has been recognised as a manifestation of DPN, reflecting disturbed proprioception and possibly abnormal muscle sensory function [4,6]. This unsteadiness has been quantified and may result in repetitive minor trauma or falls, as well as late complications including

Three classification systems for diabetic neuropathies

A. Clinical classification of diabetic neuropathies

Polyneuropathy

Sensory
- Acute sensory
- Chronic sensorimotor

Autonomic
- Cardiovascular
- Gastrointestinal
- Genitourinary
- Other

Proximal motor (amyotrophy)

Truncal

Mononeuropathy

Isolated peripheral

Mononeuritis multiplex

Truncal

B. Patterns of neuropathy in diabetes

Length-dependent diabetic polyneuropathy
- Distal symmetrical sensory polyneuropathy
- Large fibre neuropathy
- Painful symmetrical polyneuropathy
- Autonomic neuropathies

Focal and multifocal neuropathies
- Cranial neuropathies
- Limb neuropathies
- Proximal diabetic neuropathy of the lower limbs
- Truncal neuropathies

Non-diabetic neuropathies more common in diabetes
- Pressure palsies
- Acquired inflammatory demyelinating polyneuropathy

C. Classification of diabetic neuropathy

Rapidly reversible
- Hyperglycaemic neuropathy

Generalised symmetrical polyneuropathies
- Sensorimotor (chronic)
- Acute sensory
- Autonomic

Focal and mutifocal neuropathies
- Cranial
- Thoracolumbar radiculoneuropathy
- Focal limb
- Proximal motor (amyotrophy)

Superimposed chronic inflammatory demyelinating neuropathy

Figure 2.2. Three classification systems for diabetic neuropathies. Copyright © 2004 American Diabetes Association from Boulton et al. [3]. Reproduced with permission from the American Diabetes Association.

IASP pain terminology

Pain term	Definition
Allodynia	Pain caused by stimulus that does not usually provoke pain. May be static (produced by single, non-moving stimulus) or dynamic (i.e. produced by a moving stimulus)
Analgesia	Absence of pain in response to a painful stimulus that would normally be painful
Central pain	Pain initiated or caused by a primary lesion or dysfunction in the central nervous system
Dysaesthesia	An unpleasant abnormal sensation, whether spontaneous or evoked
Hyperalgesia	Increased pain response to a painful stimulus
Hyperaesthesia	Increased sensitivity to non-painful stimuli (e.g. temperature, touch)
Hyperpathia	Pain syndrome characterised by abnormally painful reaction to a stimulus, especially repetitive stimulation
Hypoalgesia	Reduced pain response to a painful stimulus
Hypoaesthesia	Decreased sensitivity to nonpainful stimuli (e.g. temperature, touch)
Neuritis	Inflammation of a nerve or nerves
Neurogenic pain	Pain initiated or caused by a primary lesion, dysfunction or transitory perturbation in the peripheral or central nervous system
Neuropathic pain	Pain initiated or caused by a primary lesion or dysfunction in the nervous system
Neuropathy	A disturbance of function or pathological change in a nerve: in one nerve, mononeuropathy; in several nerves, mononeuropathy multiplex; if diffuse and bilateral, polyneuropathy
Paraesthesia	An abnormal sensation, whether spontaneous or evoked

Figure 2.3. IASP pain terminology. Adapted with permission from Merskey H, Bogduck N. Classification of Chronic Pain: Descriptions of chronic pain syndromes and definitions of pain terms. Seattle: International Association for the Study of Pain (IASP) Press; 1994.

neuroarthropathy [6]. In most severe cases, with loss of proprioception, patients may demonstrate a positive Romberg's sign [6].

Examination of the lower limbs usually reveals a symmetrical sensory loss of vibration, pressure, pain and temperature perception (mediated by small and large fibres), and absent ankle reflexes. Sensorimotor neuropathy is often accompanied by autonomic dysfunction, signs of which include a warm or cold foot, sometimes with distended dorsal foot veins (in the absence of obstructive peripheral vascular disease), dry skin and calluses under pressure-bearing areas [1]. Any pronounced motor signs should raise the possibility of a non-diabetic aetiology, especially if asymmetrical.

Acute sensory neuropathy

Acute sensory (painful) neuropathy is a rare, distinct variety of the symmetrical polyneuropathies. It is characterised by severe sensory symptoms similar to those associated with DPN but with few neurological signs on examination. Differences between acute sensory and chronic sensorimotor neuropathies are summarised in Figure 2.4.

The overriding symptom reported by all patients is pain. This may be described as constant burning pain, discomfort (especially in the feet), severe hyperaesthesia, deep aching pain or sudden, sharp, stabbing or electric shock-like sensations in the lower limbs. Symptoms tend to be worse at night and bedclothes may irritate hyperaesthetic skin.

Other symptoms of acute sensory neuropathy include severe weight loss, depression and, in men, erectile dysfunction [3]. Clinical examination is usually relatively normal, with allodynia on sensory testing, a normal motor examination and, occasionally, reduced ankle reflexes.

Acute sensory neuropathy tends to follow periods of poor metabolic control (e.g. ketoacidosis) or a sudden change in glycaemic status (e.g. insulin neuritis), including an improvement in glycaemic control induced by oral hypoglycaemic agents. It has also been associated with weight loss and eating disorders [7].

Contrasts between acute sensory and chronic sensorimotor neuropathies

	Acute sensory	**Chronic sensorimotor**
Mode or onset	Relatively rapid	Gradual, insidious
Symptoms	Severe burning pain, aching: weight loss usual	Burning pain, paraesthesia, numbness, weight loss unusual
Symptom severity	+++	0 to ++
Signs	Mild sensory in some: motor unusual	Stocking-and-glove sensory loss: absent ankle reflexes
Other diabetic complications	Unusual	Increased prevalence
Electrophysiological investigations	May be normal or minor abnormalities	Abnormalities unusual in motor and sensory nerves
Natural history	Complete recovery within 12 months	Symptoms may persist intermittently for years: at risk of foot ulceration

Figure 2.4. Contrasts between acute sensory and chronic sensorimotor neuropathies Copyright © 2004 American Diabetes Association from Boulton et al. [3]. Reproduced with permission from the American Diabetes Association.

The natural history of acute sensory neuropathy is very different from DPN: its onset is acute or subacute but symptoms improve gradually with stabilisation of glycaemic control, and typically resolve in less than 1 year [8].

The effects of painful neuropathic symptoms on negative affect and quality of life

As noted above, the symptoms of the sensory neuropathies (both acute and chronic) vary from the extremely painful at one end of the spectrum to the painless at the other. The painful symptoms, especially the burning discomfort, electrical sensations and other uncomfortable but very difficult to describe sensory experiences, frequently cause severe physical and mental dysfunction as well as sleep disturbance [9,10], thereby negatively impacting on individuals' quality of life. The increasingly recognised symptom of unsteadiness as a consequence of impaired proprioception can also be incapacitating [10,11]. Whereas most studies, including the recent German Diabetic Microvascular Complications Study [12], have demonstrated that symptomatic neuropathy markedly diminishes patients' quality of life, such studies have invariably used generic instruments to assess health-related quality of life [13]. As the content of such instruments is imposed by the investigators and did not emerge from patients affected by neuropathic pain and other somatic experiences of neuropathy, the findings from these studies left a gap between painful neuropathy, as abstractly defined, and the patient's experience of pain, which is essential for framing effective interventions. In view of the more recent observation that only 65% of a community-based diabetes population with painful diabetic neuropathy had ever received treatment for their symptoms despite almost all reporting pain to their physician [14], the adverse effect on quality of life is likely to be even greater than previously reported. In an attempt to overcome these shortcomings, several questionnaires have recently been developed to assess quality of life from the perspective of an individual affected by diabetic neuropathy. One such measure, a neuropathy and foot ulcer-specific quality of life scale, the NeuroQoL [15], is increasingly being used in trials of new agents for the treatment of painful neuropathy in which quality-of-life measurement is an integral part of the assessment of efficacy.

Mounting evidence suggests that neuropathic symptoms have a detrimental effect on individuals' mental functioning. As an example, a cross-sectional study examined the relationship between neuropathy severity and depression, and demonstrated that neuropathic symptoms, including pain and unsteadiness, were independently associated with depressive symptomatology [9]. Moreover, this association was partially explained by two sets of psychosocial factors: (1) perceptions of symptom unpredictability and the lack of treatment control; and (2) restrictions in daily activities and diminished self-worth due to inability to perform family roles. Intriguingly, a longitudinal follow-up of this study demonstrated that only the change in neuropathic pain severity and not the baseline levels of pain predicted increased depression. This suggests that the impact of pain on depression is less likely to carry over longer periods of time [11]. By contrast, neuropathic pain appears to have an enduring association with anxiety, as demonstrated by the same research group [16]. In analyses that simultaneously examined predictors of anxiety and depression, both baseline and change in neuropathic pain predicted increases in anxiety over time.

Taken together these studies indicate that, in patients with recent increases in pain, clinicians should monitor levels of both anxiety and depression. However, anxiety should be monitored in patients even in the absence of recent increases in pain.

In view of these findings we propose that clinicians should monitor dynamics of affect and select treatment based on the predominant type of emotion, especially as commonly used medications for symptomatic neuropathy are potent mood modulators with distinct class effects on anxiety and depression. Moreover, when making the diagnosis of painful diabetic neuropathy it is important to give a full explanation of the symptoms, their causes and their treatment to the patient. Reassurance that there are no sinister underlying disorders such as malignant disease, and that the symptoms are eminently treatable and may resolve in due course, is an important part of the initial management of the patient with painful neuropathy. In contrast to the effect of painful symptoms on affect, it has also been clearly demonstrated that negative symptoms such as unsteadiness, and its consequent effects on activities of daily living, may be associated with depressive symptomatology [9]. Practitioners should therefore be aware that such negative symptoms may contribute to depressive affect that might also require treatment.

Focal and multifocal neuropathies
Mononeuropathies
Focal and multifocal neuropathies typically affect older patients with type 2 diabetes [17]. Mononeuropathies, encompassing both entrapment neuropathies and focal limb neuropathies, indicate the greater susceptibility of diabetic nerves to compression.

The median nerve is most commonly affected, giving rise to carpal tunnel syndrome. This can be demonstrated electrophysiologically in 20–30% of diabetic patients and accounts for 5.8% of all diabetic neuropathies [18]. Symptoms include painful paraesthesia of the fingers, which may progress to a deep-seated ache radiating up the forearm or, very rarely, the whole arm. This occurs primarily at night but may be precipitated during the day by repetitive wrist flexion and extension [3].

The second most common entrapment neuropathy, accounting for 2.1% of all diabetic neuropathies, results from ulnar nerve compression. It may develop as a result of deformity at the elbow joint secondary to fracture and is often associated with alcoholism. Typical symptoms include painful paraesthesiae in the fourth and fifth digits associated with hypothenar and interosseous muscle wasting [17,18].

Radial neuropathy is rare (0.6%). It presents with the characteristic motor deficit of wrist drop, occasionally accompanied by paraesthesiae in the dermatomes supplied by the superficial radial nerve. Causes include humeral fracture, blunt trauma and external compression [17,18].

Common peroneal neuropathy is the most common of all lower limb mononeuropathies. Involvement of the motor fibres in the common peroneal nerve results in weakness of the dorsiflexors and 'foot drop', whereas loss of the motor supply to the tibialis anterior muscle also leads to weakness in eversion. The resulting sensory deficit is not usually accompanied by pain or paraesthesia. Diabetes is responsible for just 10–12% of cases of peroneal neuropathy; more significant causes include external compression during anaesthesia and inappropriately placed plasters following lower-limb fractures. Once the external pressure has been relieved, most motor deficits will resolve within 3–6 months [3].

Compression of the lateral femoral cutaneous nerve is uncommon and results in pain, paraesthesia and sensory loss in the lateral aspect of the thigh (known as meralgia paraesthetica). Obesity is the most common cause, followed by trauma due to external nerve injury. Most cases resolve spontaneously [3].

Cranial neuropathies

Cranial neuropathies are extremely rare (0.05% of all diabetic neuropathies) and affect older individuals with a long duration of diabetes [19]. They primarily involve cranial nerves III, IV, VI and VII. They are thought to occur due to a microvascular infarct and typically resolve spontaneously over several months, although around 25% of patients will suffer a recurrence [3].

The classic presentation of oculomotor nerve palsy is acute-onset diplopia with ptosis and pupillary sparing associated with ipsilateral headache.

Pupillary sparing function is normal in 14–18% of patients, however, and the underlying pathology of the condition is not well understood.

Facial neuropathy, or Bell's palsy, typically presents with acute-onset unilateral weakness of facial muscles, widening of the palpebral fissure and secondary corneal irritation. This may be accompanied by taste disturbances and hyperacusis.

Very rarely, other cranial nerves may be affected in patients with diabetes. These include trigeminal neuralgia, hearing loss (cranial nerve VIII), vagal nerve involvement and vocal fold paralysis attributed to recurrent laryngeal nerve involvement.

Diabetic amyotrophy

Diabetic amyotrophy (proximal motor neuropathy) most commonly affects male type 2 diabetes patients aged 50–60 years and presents with severe pain, uni- or bilateral muscle weakness, and atrophy in the proximal thigh muscles.

Factors that contribute to the development of diabetic amyotrophy are poorly understood but may include ischaemia. The disease develops rapidly at first but then progresses more slowly over several months, indicating a combination of both vascular and metabolic factors [3]. An immune-mediated epineurial microvasculitis can present in a similar way and should be considered in the differential diagnosis – it can be demonstrated by nerve biopsy.

Diabetic truncal radiculoneuropathy

Truncal radiculoneuropathy affects middle-aged and elderly patients with diabetes and shows a predilection for men. The key symptom is pain, which is acute in onset but evolves over several months. Pain is typically described as aching or burning, may be superimposed with lancinating stabs and is worse at night in association with cutaneous hyperaesthesia.

The distribution of pain is girdle like over the lower thoracic or abdominal wall and is usually unilateral. Rarely, truncal radiculoneuropathy can result in motor weakness with bulging of the abdominal wall; symptoms may also be accompanied by profound weight loss.

Findings on neurological examination vary from no abnormalities to sensory loss and hyperaesthesia in a complete dermatomal pattern; sometimes just the ventral and dorsal rami are involved. Symptoms generally resolve within 4–6 months.

Again, the pathogenesis of diabetic truncal radiculoneuropathy is not well understood. The acute onset of symptoms suggests a vascular cause, whereas its occurrence in patients with poorer glycaemic control indicates a metabolic basis [3].

Chronic inflammatory demyelinating polyneuropathy

Chronic inflammatory demyelinating polyneuropathy (CIDP)should be considered when an unusually severe, predominantly motor neuropathy and progressive polyneuropathy develop in a diabetic patient. This diagnosis is often overlooked and the patient simply labelled as having diabetic amyotrophy or polyneuropathy, which, unlike CIDP, has no specific treatment.

By contrast, progressive symmetric or asymmetrical motor deficits, progressive sensory neuropathy in spite of optimal glycaemic control, together with typical electrophysiological findings and an unusually high cerebrospinal fluid protein level all suggest the possibility of an underlying treatable demyelinating neuropathy [20].

Autonomic neuropathy

Diabetic autonomic neuropathy [21] is a serious and common complication of diabetes that results in significant morbidity and mortality. A variation of the peripheral diabetic polyneuropathies, diabetic autonomic neuropathy can involve the entire autonomic nervous system. It may be either clinically evident or subclinical, and is manifested by dysfunction of one or more organ systems (e.g. cardiovascular, gastrointestinal, genitourinary, sudomotor, ocular) [22].

Most often, diabetic autonomic neuropathy is a system-wide disorder affecting all parts of the autonomic nervous system. Clinical symptoms generally do not arise until long after the onset of diabetes. However, subclinical autonomic dysfunction can occur within a year of diagnosis in type 2 diabetes and within 2 years in type 1 diabetes [23].

Major clinical manifestations of diabetic autonomic neuropathy include resting tachycardia, exercise intolerance, orthostatic hypertension, constipation, gastroparesis, erectile dysfunction, sudomotor dysfunction and impaired neurovascular function [3].

Because of its association with a variety of adverse outcomes, including death, cardiovascular autonomic neuropathy (CAN) is the most clinically important and well-studied form of diabetic autonomic neuropathy. It results from damage to the nerve fibres that innervate the heart and blood vessels, and results in abnormalities in heart rate control and vascular dynamics. The earliest indicator of CAN is reduced heart rate variation; later manifestations include silent myocardial ischaemia and sudden death [23].

Figure 2.5 summarises the clinical features associated with autonomic neuropathies, whereas Figure 2.6 illustrates the increase in mortality associated with CAN in diabetic individuals.

Clinical features of autonomic neuropathies

Cardiovascular
- Resting tachycardia
- Orthostatic hypotension
- Silent myocardial infarction, congestive heart failure and sudden death

Gastrointestinal
- Gastroparesis
- Diarrhoea, constipation

Genitourinary
- Bladder dysfunction
- Erectile dysfunction

Peripheral
- Gustatory sweating
- Pulpillary abnormalities

Metabolic
- Hypoglycaemia unawareness, hypoglycaemia unresponsiveness

Figure 2.5. Clinical features of autonomic neuropathies. Reproduced with permission from Vinik AI, Park TS, Stansberry KB, et al. Diabetic neuropathies. Diabetologia 2000; 43:957–973.

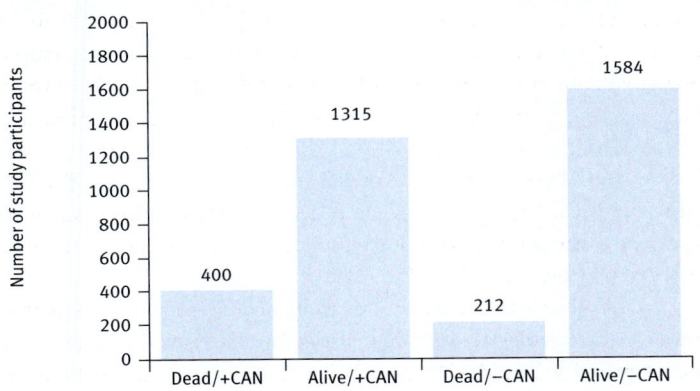

Figure 2.6. Association between CAN and mortality in diabetic patients: pooled data from 15 studies. CAN, cardiovascular autonomic neuropathy; +CAN, CAN present; –CAN, no CAN found. Copyright © 2003 American Diabetes Association from Happich et al. [12]. Reproduced with permission from the American Diabetes Association.

References

1. Boulton AJM, Vinik AI, Arezzo JC, et al. Diabetic neuropathies: a statement by the American Diabetes Association. Diabetes Care 2005; 28:956–962.
2. Partanen J, Niskanen L, Lehtinen J, et al. Natural history of peripheral neuropathy in patients with non-insulin dependent diabetes. N Engl J Med 1995; 333:89–94.
3. Boulton AJM, Malik RA, Arezzo JC, et al. Diabetic somatic neuropathies. Diabetes Care 2004; 27:1458–1486.
4. Harris MI, Eastman R, Cowie C. Symptoms of sensory neuropathy in adults with NIDDM in the U.S. population. Diabetes Care 1993; 16:1446–1452.
5. Boulton AJM, Kirsner RS, Vileikyte L. Neuropathic diabetic foot ulcers. N Engl J Med 2004; 351:48–55.
6. Van Deursen RW, Sanchez MM, Ulbrecht JS, et al. The role of muscle spindles in ankle movement perception in human subjects with diabetic neuropathy. Exp Brain Res 1998; 120:1–8.
7. Steel JM, Young RJ, Lloyd GG, et al. Clinically apparent eating disorders in young diabetic women: associations with painful neuropathies and other complications. BMJ 1987; 296:859–802.
8. Archer AG, Watkins PJ, Thomas PK, et al. The natural history of acute painful neuropathy in diabetes. J Neurol Neurosurg Psychiatry 1983; 48:491–499.
9. Vileikyte L, Leventhal H, Gonzalez JS, et al. Diabetic peripheral neuropathy and depressive symptoms: the association revisited. Diabetes Care 2005; 28: 2378–2383.
10. Gore M, Brandenburg NA, Dukes E, et al. Pain severity in diabetic peripheral neuropathy is associated with patient functioning, symptom levels of anxiety and depression, and sleep. J Pain Symptom Manage. 2005; 30:374–385.
11. Vileikyte L, Gonzalez JS, Peyrot M, et al. Predictors of depressive symptoms in patients with diabetic peripheral neuropathy: a longitudinal study. (Submitted.)
12. Happich M, John J, Stamenitis S, et al. The quality of life and economic burden of neuropathy in diabetic patients in Germany in 2002: results from the Diabetic Microvascular Complications (DIMICO) Study. Diabetes Res Clin Pract 2008; 81:223–230.
13. Vileikyte L. Psychological aspects of diabetic peripheral neuropathy. Diabetes Reviews 1999; 7:387–394
14. Daousi T, Benbow SJ, Woodward A, Macfarlane IA. A natural history of chronic painful neuropathy in a community diabetes population. Diabet Med 2006; 23:1021-1024
15. Vileikyte L, Peyrot M, Bundy C, et al. The development and validation of a neuropathy – and foot ulcer – specific quality of life instrument. Diabetes Care 2003; 26: 2549–2555.
16. Vileikyte l, Gonzalez JS, Peyrot M, et al. Differential effects of neuropathic symptoms on affect: a longitudinal study. Diabetologia. 2006; 49(suppl 1):529–530.
17. Vinik AI, Mehrabyan A, Colen L, Boulton AJM. Focal entrapment neuropathies in diabetes. Diabetes Care 2004;27:1783–1787.
18. Wilbourn AJ. Diabetic entrapment and compression neuropathies. In: Diabetic Neuropathy. Edited by PJ Dyck, PK Thomas. Philadelphia: WB Saunders, 1999; 481–508.
19. Watanabe K, Hagura R, Akanuma Y, et al. Characteristics of cranial nerve palsies in diabetic patients. Diabetes Res Clin Pract 1990; 10:19–27.
20. Ayyar DR, Sharma KR. Chronic demyelinating polyradiculoneuropathy in diabetes. Curr Diab Rep 2004; 4:409–412.
21. Freeman R. Diabetic Autonomic Neuropathy. In: Tesfaye S, Boulton AJM (eds), Oxford: Oxford University Press, Oxford Clinical Handbook of Diabetic Neuropathy. 2008 (In press).

22. Vinik AI, Maser RE, Mitchell BD, et al. Diabetic autonomic neuropathy. Diabetes Care 2003; 26:1553–1579.
23. Pfeifer MA, Weinberg CR, Cook DL, et al. Autonomic neural dysfunction in recently diagnosed diabetic subjects. Diabetes Care 1984; 7:447–453.

Chapter 3

Diagnosis and staging

Patient history

The diagnosis of diabetic neuropathy largely depends on a careful clinical examination that should be preceded by a history, which can be difficult and time-consuming. History-taking can also be confounded by the inability of some patients to communicate unusual sensations (see below) and difficulty in separating out multiple symptoms.

The patient history should include an assessment of multiple dimensions of quality of life including mood, physical and social functioning, and social support [1]. The following points should be covered [2]:
- symptom onset – when and how;
- location(s) – drawing on a map can be useful (Figure 3.1);
- radiation – this can also be expressed on a map;
- change since onset;
- characteristics/quality of pain, including use of descriptors;
- symptom severity (visual analogue scale or verbal rating scale);
- associated distress – unpleasantness of symptoms, relationship to mood;
- aggravating factors;
- alleviating factors;
- effect on function and activities of daily living;
- response to past treatments;
- psychological factors (e.g. depression, anxiety, post-traumatic stress disorder);
- coping skills (patient and family);
- smoking, alcohol and other drug use; and
- past medical history, current general health status.

A template for patient pain drawing

We need to know where you have pain:

- On the drawing below, please shade in the areas where you feel pain.
- Put an X on the area that hurts most.

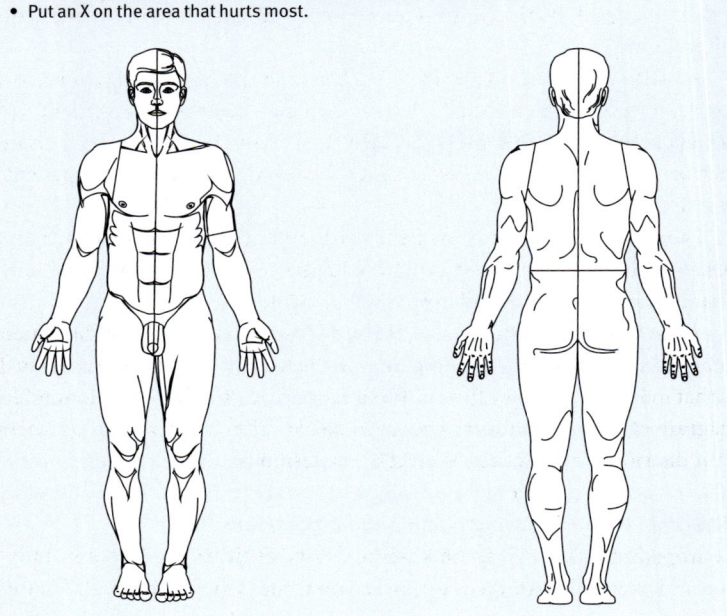

Figure 3.1. A template for patient pain drawing.

Diabetic neuropathy is a diagnosis of exclusion and the initial step is to eliminate non-diabetic causes. The main non-diabetic causes of symptomatic neuropathy are summarised in Figure 3.2.

Non-diabetic causes of symptomatic neuropathy

Malignancy (e.g. bronchogenic carcinoma)

Metabolic causes (e.g. porphyria)

Toxicity (e.g. alcohol)

Infection (e.g. HIV)

Iatrogenic (e.g. isoniazid, vinca alkaloids, chemotherapy, antiviral treatment)

Iatrogenic non-pharmacological (e.g. post-surgery)

Figure 3.2. Non-diabetic causes of symptomatic neuropathy.

Patient description of pain

It is important to emphasise the difficulties in the description and assessment of painful symptoms. Pain is a very individual sensation and patients with similar pathological lesions may describe their symptoms in markedly different ways.

As stated by Huskisson: 'Pain is a personal psychological experience and an observer can play no legitimate part in its direct management' [3]. When recording symptoms, physicians must avoid the temptation to interpret or translate patient reports; instead, they should record the patient's description verbatim.

Salient clinical features of neuropathic pain are listed in Figure 3.3. See also Chapter 2 for the International Association for the Study of Pain's definitions of pain terminology.

All neuropathic pains are projected, and the simplest way for the patient to indicate the location of their pain is by making it on a simple chart, such as that in Figure 3.1. It is often useful to relate pain distribution to a standard diagram of dermatomal and segmental distribution because, in most cases, pain distribution matches the level of the lesion [4].

Positive and negative sensory symptoms

Neuropathic sensory symptoms can be classed as either positive or negative, and it is important to distinguish between the two (Figure 3.4). Positive symptoms arise spontaneously or in response to a stimulus, whereas negative symptoms represent decreased responsiveness to stimuli (e.g. unsteadiness, numbness).

Salient clinical features of neuropathic pain

Presence of abnormal, unpleasant sensation (dysaesthesia), often with 'electric shock' or burning qualities

Pain is often paroxysmal, 'shooting' or 'stabbing'

Normally non-noxious stimulation produces pain (allodynia)

Pain may be felt in a region of sensory deficit

Pain may increase with repeated stimuli, and persist beyond removal of stimulus

No clear temporal relationship between development of pain and initial insult or injury to the nervous system

Figure 3.3. Salient clinical features of neuropathic pain. Adapted with permission from Teng JM, Mekhail NM. Neuropathic pain: mechanisms and treatment options. Pain Practice 2003; 3:8–21.

Negative and positive signs of nervous system dysfunction

Negative phenomena

Motor	Sensory	Autonomic
Weakness	Hypoaesthesia	Vasodilatation
Incoordination	Hypoalgesia	Anhidrosis
	Analgesia	Piloerection deficits
	Anosmia	
	Deafness	
	Visual loss	

Positive phenomena

Motor	Sensory	Autonomic
Fasciculations	Paraesthesia	Vasoconstriction
Dystonia	Dysaesthesia	Hyperhidrosis
Hypertonia	Allodynia	Piloerection
	Hyperalgesia	
	Hyperpathia	

Figure 3.4. Negative and positive signs of nervous system dysfunction.

It has been suggested that positive sensory symptoms should be further classified into painful and non-painful categories, as shown in Figure 3.5. This grouping is somewhat arbitrary but may be useful for certain purposes.

Descriptions of positive neuropathic sensory symptoms

Non-painful	Painful
Thick	Prickling
Stiff	Tingling
Asleep	Knife-like
Throbbing	Electric shock-like
Allodynia	Squeezing
Hyperalgesia	Constricting
	Hurting
	Burning
	Freezing
	Throbbing
	Allodynia
	Hyperalgesia

Figure 3.5. Descriptions of positive neuropathic sensory symptoms. Adapted from Apfel SC, Asbury AK, Bril V, et al. Positive neuropathic sensory symptoms as endpoints in diabetic neuropathy trials. J Neurol Sci 2001; 189:3–5.

Assessment scales

A number of questionnaires and scales have been developed to help the clinician characterise neuropathic pain with more precision; these are described on the following pages.

Visual analogue and verbal descriptor scales
Simple self-report scales, such as a 10-cm visual analogue scale or a 5-point verbal rating scale, provide unidimensional information on the severity of pain. They may be used to follow a patient's response to treatment but provide no insight into the nature or quality of pain.

Likert scales
Likert scales are another unidirectional means of assessing pain, in which patients rate their response to a question or statement on a numerical chart. For example, they may be asked to respond to a statement such as 'Pain interferes with my life on a daily basis', where 1 indicates 'strongly disagree', 5 indicates 'strongly agree' and 3 indicates 'neither agree nor disagree'.

McGill Pain Questionnaire
The McGill Pain Questionnaire (MPQ), in its original long form [5] or modified short form [6], measures a number of dimensions of pain, with 78 descriptors (15 in the short form) of sensory, affective and evaluative qualities. However, the MPQ was not designed for quantitative evaluation of distinct pain qualities and the considerable overlap between descriptors of neuropathic and nociceptive pain limits its usefulness as a diagnostic tool [4].

Brief Pain Inventory
The Brief Pain Inventory (BPI) was developed by the World Health Organization as a pain assessment tool for cancer patients [7]. It measures both the intensity of pain (sensory dimension) and the impact of pain on the patient's life (reactive dimension). It also asks the patient about pain relief, pain quality and their perception of the cause of pain. The BPI is a powerful tool and has demonstrated both reliability and validity across cultures and languages. It is used in clinical pain assessment and epidemiological studies, and to gauge the efficacy of pain treatment.

Neuropathic Pain Scale
The Neuropathic Pain Scale (NPS) was specifically designed to assess the distinct pain qualities associated with neuropathic pain [8]. It has intensity

scales (0–10) for different aspects of neuropathic pain, including 'sharp', 'hot' and 'itchy'. Most NPS items have been shown to be sensitive to treatment effects [9], and may be useful for mechanism-led diagnosis.

Leeds Assessment of Neuropathic Symptoms and Signs Pain Scale

The Leeds Assessment of Neuropathic Symptoms and Signs Pain Scale was developed to distinguish between neuropathic and non-neuropathic (nociceptive) pain [10]. It incorporates a bedside examination as well as a patient questionnaire covering four pain descriptor groups (tingling, hyperaesthesia, electric, burning). It also includes an item for skin discoloration. The examination comprises two simple bedside tests: allodynia to light touch and altered pin-prick threshold.

NeuroQoL

NeuroQoL is the first instrument to specifically assess the impact of diabetic neuropathic symptoms and deficits on patients' quality of life [11]. It includes a symptom checklist and covers restrictions in activities of daily living, problems with interpersonal relationships and changes in self-perception.

Other assessment scales

Many more assessment scales are available to record symptom quality and severity. One is a simplified neuropathy symptom score that was used in European prevalence studies and may also be useful in clinical practice [12,13]. The Michigan Neuropathy Screening Instrument is a brief 15-item questionnaire that can be administered to patients as a screening tool for neuropathy [14]. Finally, the Diabetic Neuropathy Symptom Score can be used to diagnose distal polyneuropathy [15].

Clinical examination

The most important component in the clinical diagnosis of diabetic neuropathy is the neurological examination of the lower limbs [16]. A comprehensive clinical examination is necessary to elicit sensory, motor and autonomic signs, with the aim of demonstrating abnormalities in the distribution of a nerve, plexus, root or central pathway.

Signs of nervous system dysfunction may manifest as either positive or negative phenomena, as discussed earlier. Sensory signs include increased or decreased perception of stimuli, whereas motor signs might include abnormal tone, power or coordination. These signs can be indicative of widespread central nervous system involvement. Autonomic signs tend to be localised and include changes in skin temperature or colour, swelling and sweat production.

A simple clinical examination can be carried out using easily available instruments. These might include cotton wool for assessment of light touch, a 10 g monofilament for protective sensation, a pin for pain sensation, beakers of ice/warm water for thermal sensation and a 128-Hz tuning fork for vibration sensation.

As with neuropathic symptoms, clinical signs can be assessed using composite scores such as the Neuropathic Disability Score (NDS). A modified version of the NDS has been shown to be an excellent predictor of outcomes and can be used in the community by trained non-specialists (Figure 3.6). It assigns points for the presence or absence of the following features: vibration perception threshold at the apex of the big toe; temperature perception on the dorsum of the foot; pin-prick proximal to the big toenail; and Achilles' reflex. The maximum score is 10, with a score of 6 or more indicating a high risk of foot ulcers.

The modified Neuropathic Disability Score (NDS)

	NDS	Right	Left
Vibration perception threshold 128 Hz tuning fork; apex of big toe: normal = can distinguish vibrating/not vibrating	Normal = 0 Abnormal = 1		
Temperature perception on dorsum of the foot Use tuning fork with beaker of ice/warm water			
Pin prick Apply pin proximal to big toenail just enough to deform the skin; trial pair = sharp, blunt; normal = can distinguish sharp/not sharp			
Achilles' reflex	Present = 0 Present with reinforcement = 1 Absent = 2		
NDS total out of 10			

Figure 3.6. The modified Neuropathic Disability Score (NDS). Copyright © 2004 American Diabetes Association from Boulton et al. [17]. Reproduced with permission from the American Diabetes Association.

Quantitative sensory testing

Quantitative sensory testing (QST) is an umbrella term for testing protocols that aim to give more precise determination of perception thresholds for different sensations. QST measures can be used to identify the sensory modalities affected and to estimate the magnitude of the deficit (Figure 3.7).

Among patients with diabetes, vibration, thermal and pain thresholds have proven valuable in the detection of subclinical neuropathy, in tracking the progression of disease in large cohorts and in predicting which patients are at risk of foot ulceration [17].

The strengths of QST are well documented and include the following:
- accurately controlling stimulus characteristics;
- the ability to assess multiple modalities;
- the use of well-established psychophysical procedures to enhance sensitivity;
- the capacity to measure function over a wide dynamic range of intensities, supporting the evaluation of multiple degrees of neuropathy;
- the ability to measure sensation at multiple anatomical sites, enabling the exploration of a potential distal-to-proximal gradient of sensory loss; and
- the availability of data from large, age-matched normal comparison groups.

QST also has some limitations; most importantly, it is a semi-subjective measure that is limited by the individual's attention, motivation and

Nerve fibres, function and abnormality tested by QST

Fibre	Nerve function	Mode of detection	Abnormality
C	Warm sensation Cold pain Heat pain (hairless skin) Muscle pain	Warm detection threshold Cold pain threshold Heat pain threshold Algonometer	Heat hypoaesthesia Cold allodynia Heat allodynia Static mechanical allodynia (pressure)
Aβ	Cold sensation Cold pain Heat pain (hairy skin) Pin prick	Cold detection threshold Cold pain threshold Heat pain threshold von Frey	Cold hypoaesthesia Cold allodynia Heat allodynia Static mechanical allodynia (punctate)
Aδ	Vibration	Vibration stimulus Stroke from paint brush	Dynamic mechanical allodynia

Figure 3.7. Nerve fibres, function and abnormality tested by QST. QST, quantitative sensory testing. Adapted with permission from Greenspan JD. Quantitative assessment of neuropathic pain. Curr Pain Headache Rep 2001; 5:107–113.

cooperation, as well as by anthropometric variables such as age, sex, body mass, and history of smoking and alcohol consumption [17]. In addition, QST is sensitive to changes in structure or function along the entire neuroaxis from nerve to cortex, and is not a specific measure of peripheral nerve function [18].

QST was recently evaluated by a subcommittee of the American Academy of Neurology [19]. The report concluded that QST testing is 'safe, effective, and established', and endorsed its utility in evaluating vibratory and cooling thresholds. However, the panel warned that QST should not be used as the sole criterion to define diabetic neuropathy.

Electrophysiological testing

Electrophysiological tests have emerged as important specialist tools in the evaluation of diabetic neuropathies. They are objective, parametric, non-invasive and highly reliable measures. An appropriate battery of electrophysiological tests allows the clinician to measure the speed of sensory and motor conduction, the amplitude of the propagating neural signal, the density and synchrony of muscle fibres activated by maximal nerve stimulation, and the integrity of neuromuscular transmission [17].

A key role for electrophysiological assessment is to identify neuropathies superimposed on distal symmetrical sensorimotor dysfunction. However, the tests measure only large afferent and motor nerve function , and are therefore of limited use in neuropathic pain.

Specific electrophysiological measures that can be useful in the investigation of distal sensorimotor neuropathies include the following:
- nerve conduction velocity;
- peak amplitudes, area and duration (sensory and/or motor nerves);
- F-waves;
- distribution of velocities; and
- excitability.

Diagnosing cardiovascular autonomic neuropathy

Cardiovascular autonomic neuropathy (CAN) is one of the most overlooked serious complications of diabetes and should be looked for in all patients, even in the absence of symptoms. Based on expert consensus and clinical experience, screening should be instituted at diagnosis of type 2 diabetes and 5 years after the diagnosis of type 1 diabetes. Screening should comprise a history and an examination for signs of autonomic dysfunction. The battery of tests for assessing CAN, as recommended by the American

Diagnostic tests for CAN

- **Resting heart rate**
  >100 beats/min is abnormal

- **Beat-to-beat heart rate variability**[a]
  With the patient at rest and supine (not having had coffee or a hypoglycaemic episode the night before), heart rate is monitored by ECG or autonomic instrument while the patient breathes in and out at six breaths per minute, paced by a metronome or similar device. A difference in heart rate of >15 beats/min is normal, <10 beats/min is abnormal. The lowest normal value for the expiration:inspiration ratio of the R–R interval is 1.17 in people 20–24 years of age. There is a decline in the value with age[b]

- **Heart rate response to standing**[a]
  During continuous ECG monitoring, the R–R interval is measured at beats 15 and 30 after standing. Normally, a tachycardia is followed by reflex bradycardia. The 30:15 ratio is >1.03

- **Heart rate response to the Valsalva manoeuvre**[a]
  The patient forcibly exhales into the mouthpiece of a manometer to 40 mmHg for 15 seconds during ECG monitoring. Healthy individuals develop tachycardia and peripheral vasoconstriction during strain and an overshoot bradycardia and rise in blood pressure with release. The ratio of longest R–R to shortest R–R should be >1.2

- **Systolic blood pressure response to standing**
  Systolic blood pressure is measured in the supine individual. The patient stands, and the systolic blood pressure is measured after 2 minutes. Normal response is a fall of <10 mmHg, borderline is a fall of 10–29 mmHg and abnormal is a fall of >30 mmHg with symptoms

- **Diastolic blood pressure response to isometric exercise**
  The patient squeezes a handgrip dynamometer to establish a maximum. The grip is then squeezed at 30% maximum for 5 minutes. The normal response for diastolic blood pressure is a rise of >16 mmHg in the other arm

- **ECG QT/QTc intervals**
  The QTc should be <440 ms

- **Spectral analysis**
  Very-low-frequency peak Q (sympathetic dysfunction)
  Low-frequency peak Q (sympathetic dysfunction)
  High-frequency peak Q (parasympathetic dysfunction)
  Low-frequency:high-frequency ratio Q (sympathetic imbalance)

- **Neurovascular flow**
  Using non-invasive laser Doppler measures of peripheral sympathetic responses to nociception

Figure 3.8. Diagnostic tests for CAN. [a]These can now be performed quickly (<15 minutes) in the practitioner's office using stand-alone devices that are operator friendly. [b]Lowest normal value of expiration:inspiration ratio: age 20–24 years, 1.17; 25–29, 1.15; 30–34, 1.13; 35–29, 1.12; 40–44, 1.10; 45–49, 1.08; 50–54, 1.07; 55–59, 1.06; 60–64, 1.04; 65–69, 1.03; and 70–75, 1.02. CAN, cardiovascular autonomic neuropathy; ECG, electrocardiogram. Reproduced with permission from Vinik AI, Erbas T, Pfiefer M, et al. Diabetic autonomic neuropathy. In: Inzucchi SE (ed.), The Diabetes Mellitus Manual: A primary care companion to Ellenberg and Rifkin's 6th edn. New York: McGraw Hill, 2004: 351.

Diabetes Association, is readily performed in the average clinic or hospital with widely available technology (Figure 3.8) [20].

Staging of neuropathy

The results of diagnostic and investigative tests can be amalgamated into a clinical stage. A staging system for chronic sensorimotor DPN was agreed by an international consensus panel in 1997 [21] and is shown in Figure 3.9.

The clinical stages are in general agreement with those proposed by Dyck [22] and are suitable for use in clinical practice, epidemiological studies and controlled clinical trials. Thus, 'no neuropathy' is equivalent to Dyck's stage N0 or N1a; 'clinical neuropathy' is equivalent to stage N1b, N2a or N2b; and 'late complications' is equivalent to stage N3 (Figure 3.10).

When to refer to a neurologist

Those patients with unusual features of neuropathy that might suggest a non-diabetic aetiology should be considered for a neurological referral: certainly those with asymmetrical or predominant motor signs, and in whom there is any suspicion of chronic inflammatory demyelinating polyneuropathy, should be referred. A full list of clinical features/conditions that should prompt a neurological referral can be found in Figure 3.11.

Stages of diabetic neuropathic pain

Stage of neuropathy[a]	Characteristics
No neuropathy	No symptoms or signs
Clinical neuropathy	
Chronic painful	Burning, shooting, stabbing pains with or without pins and needles, increased at night; absent sensation to several modalities; reduced/absent reflexes
Acute painful	Severe symptoms as above (hyperaesthesiae common), may follow initiation of insulin in poorly controlled diabetes, minor or absent signs
Painless with complete/ partial sensory loss	Numbness/deadness of feet or no symptoms, painless injury, reduced/ absent sensation, reduced thermal sensitivity, absent reflexes
Late complications	Foot lesions, neuropathic deformity, non-traumatic amputation

Figure 3.9. Stages of diabetic neuropathic pain. Types of diabetic neuropathy: frequent, sensorimotor symmetrical neuropathy (mostly chronic, sensory loss or pain); autonomic neuropathy (history of impotence and possibly other autonomic abnormalities); rare mononeuropathy (motor involvement, acute onset, may be painful); and diabetic amyotrophy (weakness/wasting usually of proximal lower-limb muscles). [a]Staging does not imply automatic progression to the next stage. The aim is to prevent, or at least delay, progression to the next stage. Copyright © 2004 American Diabetes Association from Boulton et al. [17]. Reproduced with permission from the American Diabetes Association.

Staging severity of diabetic polyneuropathy

N0	No objective evidence of diabetic neuropathy
N1	Asymptomatic polyneuropathy N1a: no symptoms or signs but neuropathic abnormalities N1b: test abnormalities[a] plus neuropathy impairment on neurological examination
N2	Symptomatic neuropathy N2a: symptoms, signs and test abnormality N2b: N2a plus significant ankle dorsiflexor weakness
N3	Disabling polyneuropathy

Figure 3.10. Staging severity of diabetic polyneuropathy. [a]Nerve conduction, quantitative sensory testing or autonomic test abnormalities. Adapted from Dyck [22] and Dyck PJ, Dyck PJB. Diabetic polyneuropathy: section III. In: Dyck PJ, Thomas PK (eds) Diabetic Neuropathy, 2nd edn. Philadelphia: W.B. Saunders, 1999: 255–278.

Indications for a neurological opinion in diabetic neuropathy

- Asymmetrical signs
- Predominant motor signs
- Rapid progression of signs
- Back or neck pain
- Family history of neuropathy
- Any suggestion of chronic inflammatory demyelinating polyneuropathy

Figure 3.11. Indications for a neurological opinion in diabetic neuropathy.

References

1. Haythornthwaite JA, Benrud-Larson LM. Psychological assessment and treatment of patients with neuropathic pain. Curr Pain Headache Rep 2001; 5:124–129.
2. Backonja MM, Galer BS. Pain assessment and evaluation of patients who have neuropathic pain. Neurol Clin 1998; 16:775–790.
3. Huskisson EC. Measurement of pain. Lancet 1974; ii:1127–1131.
4. Hansson P. Neuropathic pain: clinical characteristics and diagnostic workup. Eur J Pain 2002; 6(suppl A):47–50.
5. Melzack R. The McGill Pain Questionnaire: major properties and scoring methods. Pain 1975; 1:277–299.
6. Melzack R. The short-form McGill Pain Questionnaire. Pain 1987; 30:191–197.
7. Cleeland CS, Ryan KM. Pain assessment: global use of the Brief Pain Inventory. Ann Acad Med Singapore 1994; 23:129–138.
8. Galer BS, Jensen MP. Development and preliminary validation of a pain measure specific to neuropathic pain: the Neuropathic Pain Scale. Neurology 1997; 48:332–338.
9. Galer BS, Jensen MP, Ma T, et al. The lidocaine patch 5% effectively treats all neuropathic pain qualities: results of a randomized, double-blind, vehicle-controlled, 3-week efficacy study with use of the neuropathic pain scale. Clin J Pain 2002; 18:297–301.
10. Bennett M. The LANSS Pain Scale: the Leeds assessment symptoms and signs. Pain 2001; 92:147–157.
11. Vileikyte L, Peyrot M, Bundy C, et al. The development and validation of a neuropathy and foot ulcer specific quality of life rate. Diabetes Care 2003; 26:2549–2555.
12. Young MJ, Boulton AJM, MacLeod AF, et al. A multicentre study of the prevalence of diabetic peripheral neuropathy in the United Kingdom hospital clinic population. Diabetologia 1993; 36:150–154.
13. Cabezas-Cerrato J. The prevalence of clinical diabetic polyneuropathy in Spain: study in primary care and hospital clinic groups. Neuropathy Spanish Study Group of the Spanish Diabetes Society (SDS). Diabetologia 1998; 41:1263–1269.
14. Feldman EL, Stevens MJ, Thomas PK, et al. A practice two-step quantitative clinical and electrophysiological assessment for the diagnosis and staging of diabetic neuropathy. Diabetes Care 1994; 17:1281–1289.
15. Meijer JW, Smit AJ, Sondersen EV, et al. Symptom scoring systems to diagnose distal polyneuropathy in diabetes: the Diabetic Neuropathy Symptom Score. Diabet Med 2002; 19:962–965.
16. Valk GD, Nauta JJP, Strijem RLM, et al. Clinical examination versus neurophysiological examination in the diagnosis of diabetic polyneuropathy. Diabet Med 1992; 9:716–721.
17. Boulton AJM, Malik RA, Arezzo JC, et al. Diabetic somatic neuropathies. Diabetes Care 2004; 27:1458–1486.
18. Arezzo JC. Quantitative sensory testing. In: Gris FA, Cameron NE, Low PA, et al. (eds), Textbook of Diabetic Neuropathy. Stuttgart: Thième, 2003: 184–189.
19. Shy ME, Frohman EM, So YT, et al. Quantitative sensory testing: report of the Therapeutics and Technology Assessment Subcommittee of the American Academy of Neurology. Neurology 2003; 60:898–904.
20. Boulton AJM, Vinik AI, Arezzo JC, et al. Diabetic neuropathies: a statement by the American Diabetes Association. Diabetes Care 2005; 28:956–962.
21. Boulton AJM, Gries FA, Jervell JA. Guidelines for the diagnosis and outpatient management of diabetic peripheral neuropathy. Diabet Med 1998; 15:508–514.
22. Dyck PJ. Severity and staging of diabetic polyneuropathy. In: Gris FA, Cameron NE, Low PA, et al. (eds), Textbook of Diabetic Neuropathy. Stuttgart: Thième, 2003: 170–175.

Chapter 4

Management of neuropathic pain

Neuropathic pain affects many areas of the patient's life, so a multidisciplinary approach is essential in managing the condition. In many cases the clinical assessment will indicate the need for psychological or physical therapy in addition to standard pharmacological treatments. Similarly, support and information on practical measures – such as using a bed cradle to lift bedclothes off hyperaesthetic skin – often prove invaluable.

Additionally, all patients will benefit from a full explanation of their condition. This is vital for two reasons:
1. To allay fears and misconceptions and reassure patients that their symptoms may resolve in due course; and
2. To reinforce the importance of maintaining good foot care and glycaemic control in preventing serious complications such as foot ulceration and Charcot's neuroarthropathy.

The initial management of diabetic patients with symptomatic neuropathy is summarised in Figure 4.1.

Metabolic control

Several long-term studies have shown that prolonged hyperglycaemia is a major contributory factor in the development and progression of diabetic neuropathy [1–3]. In addition, there is increasing evidence that blood glucose flux (i.e. rapid changes in glycaemic control) may be important in the pathogenesis of neuropathic pain [4]. For example, a sudden improvement in metabolic control in patients started on insulin treatment may trigger a form of painful neuritis [5].

Thus, achieving near normoglycaemia and stable blood glucose control should be the first step in the management of any form of diabetic neuropathy. The optimal approach is through pancreas or islet-cell transplantation although, as most published studies involved a combined kidney and pancreas transplant, and the recipients generally had a long duration of diabetes and

> **Initial management of symptomatic neuropathy**
>
> 1. Exclude non-diabetic causes
> 2. Provide explanation, support and information on practical measures
> 3. Assess blood glucose control (regular self-monitoring and HbA1c)
> 4. Aim for optimal, stable glycaemic control
> 5. Consider pharmacological therapy

Figure 4.1. Initial management of symptomatic neuropathy.

established neuropathy, only modest improvements in neuropathic symptoms were seen [6]. However, most recently there is evidence that even in severe neuropathy, improvement in small fibre nerve function might occur after normalisation of glycaemic control [7].

Nevertheless, the benefits of achieving stable near normoglycaemia in painful neuropathy are supported by a number of small, open-label, uncontrolled studies. In one such study, patients with painful neuropathy were treated with continuous subcutaneous insulin infusion for a period of 4 months [8]. As well as relieving neuropathic pain, the intervention improved quantitative measures of nerve function. The fact that blood glucose flux was reduced in this early study might explain the symptomatic benefits of this treatment in the light of more recent observations [4]. It, therefore, appears that the stability of glycaemic control is as important as the level of achieved control.

Pharmacological treatments

There are no currently available pharmacological treatments that restore nerve function, although several disease-modifying agents are under investigation. Thus, with the exception of optimising metabolic control, the management of diabetic neuropathy is based around the treatment of symptoms. Over-the-counter analgesics may offer short-term pain relief but long-term management normally requires the prescription of systemic or topical pharmacological agents, sometimes alongside complementary treatments such as acupuncture. While offering symptomatic relief, such approaches do not alter the underlying disease process. It should be noted that, although differences exist between acute and chronic sensory neuropathies, the principles of symptomatic management are the same for both conditions. In addition, it should also be noted that the majority of the drugs commonly used are not licensed for the treatment of painful diabetic neuropathy.

Figure 4.2 summarises the dosages and side effects of the most widely prescribed pharmacological agents for painful diabetic neuropathy, while Figure 4.3 lists the number needed to treat.

Oral symptomatic therapy of painful diabetic neuropathy

Drug class	Drug	Daily dose (mg)	Side effects
Tricyclic antidepressants	Amitriptyline	25–150	++++
	Imipramine	25–150	++++
	Desipramine	25–150	++++
SSRIs	Citalopram	40	+++
	Paroxetine	40	+++
SNRIs	Duloxetine[a]	60–120	++
Anticonvulsants	Carbamazepine	200–800	+++
	Gabapentin[a]	900–3600	++
	Pregabalin[a]	150–600	++
Antiarrhythmics	Mexiletine[b]	Up to 450	+++
Opioids	Tramadol	50–400	+++
	Oxycodone CR[c]	10–60	++++

Figure 4.2. Oral symptomatic therapy of painful diabetic neuropathy.
[a]Licensed for the treatment of painful diabetic neuropathy in Europe.
[b]Mexiletine should be used with caution and with regular electrocardiogram monitoring. [c]Oxycodone controlled release (CR) may be useful as an add-on therapy in severe symptomatic neuropathy. SNRIs, serotonin and noradrenaline reuptake inhibitors; SSRIs, selective serotonin reuptake inhibitors. Adapted from McQuay et al. [9].

Number needed to treat with widely prescribed agents

Drug	Number needed to treat[a] (CI)
Carbamazepine	3.3 (2–9.4)
Duloxetine	4.3 (2.8–9.2)
Gabapentin	3.7 (2.4–8.3)
Mexiletine	10 (3–∞)
Phenytoin	2.1 (1.5–3.6)
Pregabalin	3.3 (2.3–5.9)
Tricyclic antidepressants	2.4 (2–3)
Topiramate	3.0 (2.3–4.5)
Tramadol	3.4 (2.3–6.4)

Figure 4.3. Number needed to treat with widely prescribed agents [a]Number needed to treat to achieve pain relief in one patient. CI, 95% confidence interval. Data from Max [10] and Goldstein et al. [11].

Tricyclic antidepressants

Tricyclic antidepressants, such as amitriptyline and imipramine, have been a mainstay in the treatment of painful diabetic peripheral neuropathy since the 1970s. Their efficacy is supported by several randomised controlled trials and a systematic review [9,12–14]. They have been shown to promote successful analgesia to thermal, mechanical and electrical stimuli in diabetic patients; putative mechanisms underlying these effects include inhibition of noradrenaline and/or serotonin reuptake at synapses of central descending pain control systems and, more recently, the antagonism of N-methyl-D-aspartate (NMDA) receptors, which mediate hyperalgesia and allodynia.

Most clinical experience of tricyclic antidepressants involves amitriptyline or imipramine. For both agents the dosage required for symptomatic relief is 25–150 mg/day, although in older patients it can be useful to start at 10 mg/day. To avoid undue drowsiness, the dose can be taken once daily, usually in the evening or at bedtime. Despite their proven efficacy, the major drawback of tricyclic drugs is the frequency of side effects related mainly to their anticholinergic properties. These include sedation, blurred vision, dry mouth, orthostatic hypertension and cardiac arrhythmias. Of patients prescribed tricyclic antidepressants for the relief of pain, 7–58% report either ineffective pain relief or side effects precluding optimal dosage [10,15].

Selective serotonin reuptake inhibitors

Selective serotonin reuptake inhibitors (SSRIs, e.g. paroxetine, citalopram) are specific for the presynaptic reuptake of serotonin but not noradrenaline, and thus have fewer side effects than tricyclic antidepressants. Their mode of action is based around experimental observations that serotonin is an important mediator of analgesia. In placebo-controlled trials, paroxetine and citalopram, both at 40 mg/day, have been shown to be effective in relieving neuropathic pain [16,17]. However, they are less effective than tricyclic antidepressants.

Serotonin and noradrenaline reuptake inhibitors

The serotonin and noradrenaline reuptake inhibitor (SNRI) duloxetine has both analgesic and antidepressant effects and can be used for the treatment of diabetic peripheral neuropathic pain. In a placebo-controlled study, duloxetine 60–120 mg/day was shown to be effective in relieving neuropathic pain [11]. It significantly improved the mean 24-hour pain severity score and increased the proportion of patients with at least a 30% reduction in pain score compared with baseline. It also separated from placebo on nearly all the secondary measures including health-related outcome measures. Furthermore, duloxetine 120 mg/day seemed to work best on pain described as shooting, stabbing, sharp, hot–burning and splitting. Although 120 mg/day did not produce a significantly greater improvement compared with 60 mg/day, duloxetine treatment at 60–120 mg/day was considered to be well tolerated with less than 20% discontinuation due to adverse events, such as somnolence, dizziness or constipation. Unlike tricyclics, some anticonvulsants and opioids, it does not generally require dose titration. More recently, an analysis of three randomised controlled trials of duloxetine in the management of neuropathic pain in diabetes confirmed that the drug is efficacious and well tolerated [18].

Another similar drug in this class, bicifadine, is currently under investigation in painful diabetic neuropathy [19].

Anticonvulsants

Anticonvulsants (e.g. gabapentin, carbamazepine, phenytoin, lamotrigine, topiramate and pregabalin) have been used in the management of neuropathic pain for many years. The evidence for the efficacy of first-generation agents (e.g. carbamazepine, phenytoin) is limited, coming mainly from small, single-centre studies [20–22]. Furthermore, these agents are associated with a relatively high frequency of adverse events, particularly central effects such as somnolence and dizziness [15].

More recently, however, a number of newer agents have been found to be effective in randomised clinical trials, and these are the subject of a recent review [23]. For example, gabapentin is now widely used for the relief of neuropathic pain. It is structurally related to the neurotransmitter γ-aminobutyric acid (GABA) and was first used as an antiepileptic agent. Two large placebo-controlled trials have shown that gabapentin 1800 mg/day offers significant pain relief in diabetic neuropathy, together with reduced sleep disturbance [24,25]. Although there is no need for blood monitoring, some patients may experience minor side effects such as dizziness, somnolence, headache, diarrhoea, confusion and nausea [25]. In addition, a recent review concluded that gabapentin at dosages of 1800–3600 mg/day was effective for neuropathic pain and had a superior side-effect profile to that of the tricyclic agents [26]. Although clinical experience suggests that many patients derive some relief from slightly lower dosages, a common failure of physicians is to give an insufficient dosage for pain relief: 32% of prescriptions written for gabapentin are for less than the 900 mg starting dose, 63% are for less than 1200 mg, with only 28% of prescriptions for 1800 mg or more [27].

A second GABA analogue, pregabalin, has recently been shown to be effective in the treatment of painful diabetic neuropathy in several randomised controlled trials [28,29,30]. Based on these and other evidence supporting its efficacy and tolerability, doses of 150–600 mg/day for the treatment of diabetic neuropathic pain are recommended. However, patients often achieve satisfactory pain control without the need for the highest dose. Pregabalin, although similar structurally to gabapentin, is three times more potent and only needs to be taken twice daily, compared with gabapentin which is usually taken three times daily [31]. These properties mean that pregabalin may be easier to prescribe and is effective across the dose range, resulting in a decreased need for titration [30]. Moreover, a recent meta-analysis of seven randomised controlled trials in painful diabetic neuropathy has confirmed the efficacy and safety of pregabalin [32].

Although the most convincing evidence exists for gabapentin and pregabalin, other anticonvulsant drugs that have shown efficacy in relieving neuropathic pain in controlled clinical trials include oxcarbamazepine (now withdrawn from further study), sodium valproate, lamotrigine and topiramate [33–36]. Finally, the sodium channel blocker lacosamide [37] appears promising and is currently in phase III trials [38].

Local anaesthetic/antiarrythmic agents

The benefit of intravenous lidocaine (5 mg/kg over 30 minutes) in painful diabetic neuropathy was confirmed in a randomised, double-blind, placebo-controlled trial [39]. However, oral dosing is unavailable and electrocardiogram (ECG) monitoring is necessary during administration.

Lidocaine thus tends to be reserved for short-term use in patients with excruciating neuropathic pain.

Mexiletine is a class 1B antiarrhythmic agent and is an orally available structural analogue of lidocaine. Its efficacy in reducing neuropathic pain at doses of 225–675 mg/day has been confirmed in several clinical trials [40,41].

However, a review of seven controlled trials suggested that mexiletine provided only a modest analgesic effect [42]. Regular ECG monitoring is necessary and the long-term use of mexiletine cannot be recommended.

Opioids

Many physicians are reluctant to prescribe opioids for neuropathic pain, probably because of fears of addiction. However, as stated by Foley in an editorial, 'we must focus urgent attention on the needs of suffering patients' [43]. In particular, opioids and opioid-like drugs are often effective in treating neuropathic pain that has failed to respond to therapies such as those described above [43].

Of the orally administered opioids, tramadol is the best studied. It is a centrally acting synthetic non-narcotic analgesic with an unusual mode of action, working on both opioid and monoaminergic pathways. It has a lower abuse potential than conventional strong opioids and development of tolerance is uncommon. In a randomised controlled trial, tramadol 210 mg/day was effective in the management of painful diabetic neuropathy [44]. Although this study only lasted 6 weeks, a follow-up study suggested that symptomatic relief was maintained for at least 6 months [45]. Side effects, however, are relatively common and similar to those of other opioid-like drugs.

Finally, two randomised trials have confirmed the efficacy of controlled-release oxycodone for neuropathic pain in diabetes [46,47]. The drug was well tolerated, and oxycodone and other opioids may be considered add-on therapies for patients failing to respond to standard medications.

Unfortunately, no studies have examined the long-term use of opioids, so the risks of tolerance and dependence have yet to be quantified. In recognition of these potential problems physicians should be alert to signs of abuse; these might include claims of lost medication, premature prescription requests, intoxication or frequently missed appointments [48].

Topical treatments

Topical agents offer several theoretical advantages, including minimal systemic side effects, lack of drug interactions and no need for dose titration.

However, few have been evaluated in well-designed, randomised controlled trials.

Capsaicin

Capsaicin is an alkaloid extract of hot chilli peppers that is thought to act by depleting substance P from the terminals of unmyelinated C fibres. Several controlled studies have investigated topically applied capsaicin cream (0.075%) in the treatment of painful diabetic neuropathy, and a meta-analysis suggested overall efficacy [49]. However, a more recent trial failed to demonstrate any pain relief with this agent [50], and study blinding is compromised by the transient local hyperalgesia (usually a mild burning sensation) that occurs in many patients. This side effect, and the fact that the cream must be applied four times daily over the entire painful area, may limit the clinical usefulness of the drug.

Lidocaine patches

Lidocaine patches are approved by the US Food and Drug Administration and the European Agency for the Evaluation of Medicinal Products for the treatment of post-herpetic neuralgia only and have shown promising efficacy in painful diabetic neuropathy. In an open-label pilot study, treatment with a 5% lidocaine patch for 3 weeks provided a significant improvement in pain and quality of life [51]. However, these observations need to be confirmed in a properly designed randomised controlled trial before the patch can be recommended.

Isosorbide dinitrate spray

Isosorbide dinitrate spray applied to the lower limbs was shown to be effective in the relief of painful diabetic neuropathy in a randomised crossover trial [52]. More recently, glyceryl trinitrate patches were shown to be effective [53]. Both studies were small and single centre, however, and these topical approaches await evaluation in multicentre trials.

Non-pharmacological treatments

Acupuncture

A number of uncontrolled studies have reported significant benefits of acupuncture in the relief of neuropathic pain. In the most recent published report, up to six courses of traditional Chinese acupuncture given over a 10-week period resulted in 77% of patients reporting significant pain relief. Furthermore, during follow-up of up to 1 year, the majority of patients were able to stop or significantly reduce their other pain medication [54]. Although controlled trials are needed to confirm the efficacy of acupuncture, these are difficult to design because of the problems encountered with finding the correct site for 'sham' treatment.

Electrical therapy

Despite encouraging preliminary data, a recent randomised crossover study of pulsed electrical stimulation failed to show efficacy in the treatment of painful diabetic neuropathy [55]. A number of other physical therapies have shown promise in small controlled trials: these include low-intensity laser therapy [56], percutaneous electrical nerve stimulation [57], static magnetic field therapy [58] and monochromatic infrared light treatment [59]. However, other studies have been less encouraging, emphasising the need for multiple confirmatory randomised trials to allow an evidence-based treatment decision to be arrived at.

A case series of patients with severe painful neuropathy unresponsive to conventional treatments suggested that an implanted spinal cord stimulator was effective [60]. Although a recent follow-up study suggested that this approach might offer long-term symptomatic relief [61], electrical spinal cord stimulation cannot be recommended except in very resistant cases, because it is invasive, expensive and unproven in controlled studies.

Treatment of cardiovascular autonomic neuropathy

As noted previously, patients with diabetes should be screened for autonomic neuropathy at regular intervals (see Chapter 3). If screening is negative, then tests should be repeated annually; if positive, appropriate diagnostic tests and symptomatic treatments should be instituted. These are summarised in Figure 4.4.

The most recent position statement of the American Diabetes Association on the management of the diabetic neuropathies was published in 2005 [62].

Treatment of autonomic neuropathy

Symptoms	Tests	Treatments
Cardiac		
Exercise intolerance, early fatigue and weakness with exercise	HRV, MUGA thallium scan, ^{123}I-MIBG scan	Graded supervised exercise, ACE inhibitors, β blockers
Postural hypotension, dizziness, weakness, lightheadedness, fatigue, syncope	HRV, measure blood pressure standing and supine, measure catecholamines	Mechanical measures, clonidine, midodrine, octreotide
Gastrointestinal		
Gastroparesis, erratic glucose control	Gastric emptying study, barium study	Frequent small meals, prokinetic agents (metoclopramide, domperidone, erythromycin)
Abdominal pain or discomfort, early satiety, nausea, vomiting, belching, bloating	Endoscopy, manometry, electrogastrogram	Antibiotics, antiemetics (phenergan, compazine, tigan, hyoscine), bulking agents, tricyclic antidepressants, pancreatic extracts, pyloric botulinum toxin, gastric pacing, enteral feeding
Constipation	Endoscopy	High-fibre diet and bulking agents, osmotic laxatives, lubricating agents and prokinetic agents used cautiously
Diarrhoea, often nocturnal, alternating with constipation and incontinence		Trials of soluble fibre, gluten and lactose restriction, anticholinergic agents, cholestyramine, antibiotics, clonidine, somatostatin, pancreatic enzyme supplements
Sexual dysfunction		
Erectile dysfunction	History and physical examination, HRV penile-brachial pressure index, nocturnal penile tumescence	Sex therapy, psychological counselling, PDE5 inhibitors, prostaglandin E_1 injection, device or prosthesis
Vaginal dryness		Vaginal lubricants

Bladder dysfunction

Frequency, urgency, nocturia, urinary retention, incontinence	Cystometrogram, postvoiding, sonography	Bethanechol, Intermittent catheterisation

Sudomotor (sweating) dysfunction

Anhidrosis, heat intolerance, dry skin, hyperhidrosis	Quantitative sudomotor axon reflex, sweat test, skin blood flow	Emollients and skin lubricants, hyoscine, glycopyrrolate, botulinum toxin, vasodilators

Pupillomotor

Visual blurring, impaired adaptation to ambient light, impaired visceral sensation	Pupillometry, HRV	Care with driving at night, recognition of unusual presentations of myocardial infarction

Figure 4.4. Treatment of autonomic neuropathy. ACE, angiotensin-converting enzyme; HRV, heart rate variation; MIBG, metaiodobenzylguanidine; MUGA, multigated angiography. PDE5, phosphodiesterase type 5. Copyright © 2005 American Diabetes Association from Zinman et al. [56]. Reproduced with permission from the American Diabetes Association.

References

1. DCCT Research Group. The effect of intensive diabetes therapy on thedevelopment and progression of neuropathy. Ann Intern Med 1995; 122:561–568.
2. Partanen J, Niskanen L, Lehtinen J, et al. Natural history of peripheral neuropathyin patients with non-insulin dependent diabetes. N Engl J Med 1995;333:89–94.
3. Tesfaye S, Stevens LK, Stephenson JM, et al. Prevalence of diabetic peripheralneuropathy and its relation to glycaemic control and potential risks: the EuroDiab IDDM complication study. Diabetologia 1996; 39:1377–1386.
4. Oyibo S, Prasad YD, Jackson NJ, et al. The relationship between blood glucose excursions and painful diabetic peripheral neuropathy: a pilot study. Diabet Med 2002; 19:870–873.
5. Kihara M, Zollman PJ, Smithson IL, et al. Hypoxic effect of exogenous insulin on normal and diabetic peripheral nerve. Am J Physiol 1994; 226(6Pt 1):E980–E985.
6. Kennedy WR, Navarro X, Goetz FC, et al. Effects of pancreas transplantation on diabetic neuropathy. N Engl J Med 1990; 322:1031–1037.
7. Mehra S, Tavakoli M, Kallinikos PA, et al. Corneal confocal microscopy detects early nerve regeneration after pancreas transplantation in patients with type 1 diabetes. Diabetes Care 2007;30:2608–2612
8. Boulton AJM, Drury J, Clarke B, et al. Continuous subcutaneous insulin infusion in the management of painful diabetic neuropathy. Diabetes Care 1982;5:386–390.
9. McQuay H, Tramer M, Nye BA. A systemic review of antidepressants in neuropathic pain. Pain 1996: 68:217–227.
10. Max MB. Antidepressants as analgesics. In: Fields HL, Liebeskind JC (eds), Pharmacological approaches to the treatment of chronic pain: new concepts and critical issues. Progress in Pain Research and Management (Vol 1). Seattle: IASP Press,1994: 229–246.
11. Goldstein DJ, Lu Y, Detke MJ, et al. Duloxetine vs placebo in patients with diabetic neuropathy. Pain 2005; 116:109–118.
12. Max MB, Lynch SA, Muir J, et al. Effects of desipramine, amitriptyline, and fluoxetine on pain in diabetic neuropathy. N Engl J Med 1992; 326:1250–1256.
13. Watson CP. The treatment of neuropathic pain: antidepressants and opioids.Clin J Pain 2000; 16(2 suppl):S49–S55.
14. Mendell JR, Sahenk Z. Painful sensory neuropathy. N Engl J Med 2003;348:1243–1255.
15. Boulton AJM, Malik RA, Arezzo JS, et al. Diabetic somatic neuropathies. Diabetes Care 2004; 27:1458–1486.
16. Sindrup SH, Gram LF, Brosen K, et al. The SSRI paroxetine is effective in the treatment of diabetic neuropathy symptoms. Pain 1990;42:135–144.
17. Sindrup SH, Bjerre U, Dejgaard A, et al. The selective serotonin reuptake inhibitor citalopram relieves the symptoms of diabetic neuropathy. Clin Pharmacol Ther 1992; 52:547–552.
18. Kajdasz DK, Iyengar S, Desaiah D, et al. Duloxetine for the management of diabetic peripheral neuropathic pain: evidence-based findings from post hoc analysis of three multicenter, randomized, double-blind, placebo-controlled, parallel-group studies. Clin Ther 2007;29(suppl 2):536–546.
19. Basile AS, Janowsky A, Golembiowska K, et al. Characterization of the antinociceptive actions of bicifadine in models of acute, persistent, and chronic pain. J Pharmacol Exp Ther 2007;321:1208–1225.
20. Rull JA, Quibrera R, Gonzalez-Millan H, et al. Symptomatic treatment of peripheral diabetic neuropathy with carbamazepine (Tegretol): double-blind crossover trial. Diabetologia 1969; 5:215–218.
21. Saudek CD, Werns S, Reidenberg MM. Phenytoin in the treatment of diabetic symmetrical polyneuropathy. Clin Pharmacol Ther 1977; 22:196–199.

22. Kochar DK, Jain N, Agarwal RP, et al. Sodium valproate in the management of painful neuropathy in type 2 diabetes – a randomized placebo controlled study. Acta Neurol Scand 2002; 106:248–252.
23. Vinik AI. Use of antiepileptic drugs in the treatment of chronic painful diabetic neuropathy. J Clin Endocrinol Metab 2005; 90:4936–4945.
24. Backonja MM, Beydoun A, Edwards KR, et al. Gabapentin for symptomatic treatment of painful neuropathy in patients with diabetes mellitus. JAMA 1998;280:1831–1836.
25. Rowbotham M, Harden N, Stacey B, et al. The gabapentin study group. Gabapentin for the treatment of postherpetic neuralgia. JAMA 1998; 280:1837–1843.
26. Backonja M, Glazman RI. Gabapentin dosing for neuropathic pain evidence from randomized placebo controlled clinical trials. Clin Ther 2003; 25:81–104.
27. DIN-LINK data, Cegedim Strategic Data, October 2006. Data calculated based on prescriptions where dosing instructions are known.
28. Rosenstock J, Tuchman M. LaMoreau L, et al. Pregabalin for the treatment of painful diabetic neuropathy: a randomized, controlled trial. Pain 2004;110:628–634.
29. Richter RW, Portenoy R, Sharma M, et al. Relief of painful diabetic neuropathy with pregabalin: a randomized, placebo-controlled trial. J Pain 2005; 6:253–260.
30. Lesser H, Sharma U, LaMoreaux L, et al. Pregabalin relieves symptoms of painful diabetic neuropathy: a randomized controlled trial. Neurology 2004;63:2104–2110.
31. Serpell M. Recognising and treating neuropathic pain. Available at www.progressnp.com. Last accessed August 2007.
32. Freeman R, Durso-Decruz E, Emir B. Efficacy, safety, and tolerability of pregabalin treatment for painful diabetic peripheral neuropathy: findings from seven randomized, controlled trials across a range of doses. Diabetes Care 2008;31:1448–1454.
33. Beydoun A, Kobetz SA, Carrazana EJ. Efficacy of oxcarbazepine in the treatment of diabetic neuropathy. Clin J Pain 2004; 20:174–178.
34. Kochar DK, Rawat N, Agrawal RP, et al. Sodium valproate for painful diabetic neuropathy: a randomized double-blind trial. Q J Med 2004; 97:33–38.
35. Eisenberg E, Lurie Y, Braker C, et al. Lamotrigine reduces painful diabetic neuropathy. Neurology 2001; 57:505–509.
36. Edwards K, Glantz MJ, Button J, et al. Efficacy and safety of topiramate in the treatment of painful diabetic neuropathy: a double-blind placebo-controlled study. Neurology 2000; 54(suppl 3):A81.
37. Rauck RL, Shaibani A, Biton V, et al. Lacosamide in painful diabetic peripheral neuropathy: a phase 2 double-blind placebo-controlled study. Clin J Pain 2007;23:150–158.
38. Ziegler D. Painful diabetic neuropathy: treatment and future aspects. Diabet Metab Res Rev 2008;24(suppl 1):S52–S57.
39. Kastrup J, Petersen P, Dejard A, et al. Intravenous lidocaine infusion: a new treatment of chronic painful diabetic neuropathy. Pain 1987; 28:69–75.
40. Dejgard A, Peterson P, Kastrup J. Mexiletine for treatment of chronic painful diabetic neuropathy. Lancet 1988; i:9–11.
41. Oskarsson P, Lins PE, Ljunggren JG, et al. Efficacy and safety of mexiletine in the treatment of painful diabetic neuropathy. Diabetes Care 1997; 20:1594–1597.
42. Jarvis B, Coukell AJ. Mexilitene: a review of its therapeutic use in painful diabetic neuropathy. Drugs 1998; 56:691–708.
43. Foley K. Opioids and chronic neuropathic pain. N Engl J Med 2003; 348:1279–1281.
44. Harati Y, Gooch C, Swenson M, et al. Double-blind randomized trial of tramadol for the treatment of the pain of diabetic neuropathy. Neurology 1998;50:1842–1846.
45. Harati Y, Gooch C, Swenson M, et al. Maintenance of the long-term effectiveness of tramadol in treatment of the pain of diabetic neuropathy. J Diabetes Complications 2000; 14:65–70.
46. Gimbel JS, Richards P, Portenoy RK. Controlled-release oxycodone for pain in diabetic neuropathy: a randomized controlled trial. Neurology 2003; 60:927–934.

47. Watson CPN, Moulin D, Watt-Watson J, et al. Controlled-release oxycodone relieves neuropathic pain: a randomized controlled trial in painful diabetic neuropathy. Pain 2003; 105:71–78.
48. The Pain Society. Provisional Recommendations for the Appropriate Use of Opioids in Patients with Chronic Non-cancer Related Pain. London: The Pain Society, 2003.
49. Zhang WY, Li Wan Po A. The effectiveness of topically applied capsaicin: a meta analysis. Eur J Clin Pharm 1994; 46:517–522.
50. Low PA, Opfer-Gehrking TL, Dyck PJ, et al. Double-blind placebo-controlled study of capsaicin cream in chronic distal painful polyneuropathy. Pain 1995;62:163–168.
51. Barbano RL, Herrman DN, Hart-Gouleau S, et al. Effectiveness, tolerability and impact on quality of life of the 5% lidocaine patch in diabetic polyneuropathy. Arch Neurol 2004; 61:914–918.
52. Yuen KC, Baker NR, Rayman G. Treatment of chronic painful diabetic neuropathy with isosorbide dinitrate spray: a double-blind placebo-controlled cross-over study. Diabetes Care 2002; 25:1699–1703.
53. Rayman G, Baker NR, Krishnan ST. Glyceryl trinitrate patches as an alternative to isosorbide dinitrate spray in the treatment of painful neuropathy. Diabetes Care 2003; 26:2697–2698.
54. Abuaisha BB, Constanzi JB, Boulton AJM. Acupuncture for the treatment of chronic painful diabetic neuropathy: a long-term study. Diabetes Res Clin Pract 1998; 39:115–121.
55. Oyibo S, Breislin K, Boulton AJM. Electrical stimulation therapy through stocking electrodes for painful diabetic neuropathy: a double-blind controlled crossover study. Diabet Med 2004; 21:940–944.
56. Zinman LH, Ngo M, Ng ET, et al. Low-intensity laser therapy for painful symptoms of diabetic sensorimotor polyneuropathy: a controlled trial. Diabetes Care 2004; 27:921–924.
57. Hamza MA, White PF, Craig WF, et al. Percutaneous electrical nerve stimulation: a novel analgesic therapy for diabetic neuropathic pain. Diabetes Care 2000;23:365–370.
58. Weintraub MI, Wolfe GI, Baronh RA, et al. Static magnetic field therapy for symptomatic diabetic neuropathy: a randomized, double-blind, placebo controlled trial. Arch Phys Med Rehabil 2003; 84:736–746.
59. Leonard DR, Farooqu MH, Myers S. Restoration of sensation, reduced pain, and improved balance in subjects with diabetic peripheral neuropathy: a double blind, randomized placebo-controlled study with monochromatic infrared treatment. Diabetes Care 2004; 27:168–172.
60. Tesfaye S, Watt J, Benbow SJ, et al. Electrical spinal-cord stimulation for painful diabetic peripheral neuropathy. Lancet 1996; 348:1698–1701.
61. Daousi C, Benbow SJ, Macfarlane IA. Electrical spinal cord stimulation in the long-term treatment of chronic painful diabetic neuropathy. Diabet Med 2005; 22:393–398.
62. Boulton AJ, Vinik AI, Arezzo JC, et al. Diabetic neuropathies: a statement by the American Diabetes Association. Diabetes Care 2005; 28:956–962.

Index

Page references to figures are shown in *italics*.

acupuncture 42–43
acute sensory neuropathy
  incidence 11
  symptoms 11–12
  vs. chronic sensorimotor neuropathy 11
aetiology 7
affect 12–13
algorithm, pain management 36
allodynia *10*, 11, *23*, *24*, *26*, *28*
amitriptyline *37*, 38
anticonvulsants 39–40
anaesthetics 40–41
antiarrhythmics 40–41
assessment scales 25–26
autonomic neuropathy 16–18

Bell's palsy 15
Brief Pain Inventory (BPI) 25

CAN see cardiovascular autonomic neuropathy (CAN)
capsaicin 42
carbamazepine *37*, 39
cardiovascular autonomic neuropathy (CAN) 16, *17*
  diagnostic tests 29–31, *29*, *30*
  management 43, *44*
  mortality *17*
carpal tunnel syndrome 13
chronic hyperglycaemia 2
chronic inflammatory demyelinating polyneuropathy 9, 16, 31, *32*
chronic sensorimotor distal symmetric polyneuropathy 8–10
citalopram *37*, 38
classification systems 7–8, *9*
clinical examination 26–27
cranial neuropathies 14–15

desipramine *37*
Diabetes Control and Complications Trial 2–3
diabetic amyotrophy 15
diabetic autonomic neuropathy 16–17
  clinical features *17*
diabetic neuropathy
  definitions 1–2
  epidemiology 2–3
Diabetic Neuropathy Symptom Score 26
diabetic truncal radiculoneuropathy 15–16

distal symmetrical polyneuropathy (DPN)
  incidence 8
  sequelae 8
  staging 31, *31*
  symptoms 8–10
duloxetine *37*, 38–39
Dyck's staging 31

electrical therapy 43
electrophysiological testing 29
entrapment neuropathies 13–14

facial neuropathy 15

gabapentin *37*, 39–40
γ-aminobutyric acid (GABA) 39–40
glyceryl trinitrate patches 42

imipramine *37*, 38
impaired glucose tolerance (IGT) 2–3
  incidence of neuropathies 2
isosorbide dinitrate spray 42

lacosamide 40
lamotrigine 39–40
Leeds Assessment of Neuropathic Symptoms and Signs Pain Scale 26
lidocaine patches 42
lidocaine 40
Likert scales 25
local anaesthetic/antiarrythmic agents 40–41

McGill Pain Questionnaire (MPQ) 25
mexiletine *37*, 40–41

Michigan Neuropathy Screening Instrument 26
mononeuropathies 13–14

natural history of neuropathies 3
NDS *see* Neuropathic Disability Score
Neuropathic Disability Score (NDS), modified 27, *27*
neuropathic pain
  clinical features *23*
  nondiabetic causes *22*
    see also pain; pain management
  template for pain drawing *22*
Neuropathic Pain Scale (NPS) 25–26
NeuroQoL 12, 26
N-methyl-D-aspartate (NMDA) receptor antagonists 38
NPS *see* Neuropathic Pain Scale

oculomotor nerve palsy 14–15
opioids 41
oral symptomatic therapy *37*
oxcarbamazepine 40
oxycodone *37*, 41

pain
  patient description of *23*
  sensory symptoms 23–24, *24*
  template for pain drawing *22*
  terminology *10*
  see also neuropathic pain
pain management
  cardiovascular autonomic neuropathy 43, *44–45*
  metabolic control 35–36
  nonpharmacological treatments 42–43

pharmacological treatments 36–42
paroxetine *37*, 38
patient
  description of pain 23
  history 21
  pain drawing template *22*
peroneal neuropathy 14
phenytoin *37*, 39
pregabalin *37*, 39–40
prescribed agents, number needed to treat *37*
prevalence of neuropathies 2
proximal motor neuropathy 15

quality of life 12–13
quantitative sensory testing (QST) *28*, 28–29
  nerve fibres, function and abnormality 29

radial neuropathy 14
risk factors for neuropathies 2–3

selective serotonin reuptake inhibitors (SSRIs) *37*, 38
serotonin and noradrenaline reuptake inhibitors (SNRIs) *37*, 38–39
sodium valproate 40
staging 31, *31, 32*
studies *see* trials and studies

template for pain drawing *22*
tramadol *37,41*
topiramate *37*, 39–40
trials and studies
  Diabetes Control and Complications Trial 2
  UK Prospective Diabetes Study 3
  EuroDiab 3
tricyclic antidepressants *37*, 38

UK Prospective Diabetes Study 3
ulnar nerve compression 14

verbal descriptor scale 25
visual analogue scale 25